Kumsheen Journal of Anatomy and Physiology
Volume 1, Issue 1

© 2018 School District 74 – Gold Trail, All Rights Reserved.

ISBN: 978-1-387-84579-8

The Biology 12 class would like to acknowledge the support of School District 74 – Gold Trail in helping facilitate this journal. We would like to express particular gratitude to Brendan Bogle, the District Experiential Learning Specialist, for providing the resources and funding necessary to make this project a reality.

Table of Contents

Preface

    3   About the Authors

    4   Rationale

Articles

5	**Biological Aging Rates of Humans** Domonique Samson-Hayden	26	**Iron Deficiency** Nathanial Brown
7	**Creating GMO Humans** Ara Michell	29	**Human Cloning** McKenna Adams-James
10	**Junk DNA** Steven Arnouse	31	**Function of Taurine in Energy Drinks** Sadie Drynock
12	**Causes of Acne** Jacqueline Webster	33	**Importance of Vitamin C** Farah Abbott
15	**Benefits of Short-Term Fasting** Chase Johnny	35	**Alkaline and Acidic Diets** Brianne Duncan
18	**Genetic Basis of Races** Patashi Pimms	39	**Skin Graft Rejection** Patrick Maw
20	**Fetal Alcohol Exposure** Brianna Smith	41	**Reversal of the Aging Process** CJ James
22	**Effect of GMOs on Human Cell Biology** Mason Jory		

About the Authors

The Kumsheen Journal of Anatomy and Physiology is a self-published science journal containing inquiry articles that were researched and authored by the Biology 12 class of Kumsheen Secondary School. Volume 1 Issue 1 consists of articles authored by students in grade 11 and grade 12 during the 2017/18 academic year.

Biology 12 is an elective senior science course, and student authors were enrolled in the course for a variety of reasons. For many of the authors, Biology 12 is taken purely as a graduation requirement. For others, Biology 12 is a stepping stone on the path to their desired post-secondary program. Indeed, some of the authors in this journal will go on to university biology programs in the 2018/19 academic year.

May all of our student authors find success in their future pursuits, and may this journal serve as a legacy of their accomplishments and learning at Kumsheen Secondary School.

Articles were proofread and edited for format and style continuity by Mat Houlton, teacher of the Biology 12 class.

Rationale

Why perform student inquiry?

Education is changing rapidly in British Columbia. Careers are becoming more modernized, higher education is becoming more competitive, and therefore K-12 education must adapt to develop the citizens of tomorrow. This evolution is reflected emphatically in B.C.'s new curriculum. The education system is moving away from content-based learning and towards competency-based learning. "Competency-based learning" is just a fancy way of saying that we need to start focusing more on *how* students learn, and less on *what* students learn.

What does competency-based learning look like in a course as content heavy and technical as Biology 12? Of course, students still need to be building a strong biology foundation. There are still content-based learning outcomes in the new Biology 12 curriculum, and students develop a deep understanding of concepts related to homeostasis, cytology, and organ systems. The change comes in emphasizing the skills required to access future knowledge in biology. These skills include critical thinking, personal and social awareness, and communication.

Inquiry weaves in all of the core competencies of a successful learner. Students have to be personally and socially reflective in crafting questions that are meaningful to them. Students have to think critically about the information they find. Students ultimately must be able to communicate their knowledge and understanding effectively. These skills do more than just help students meet learning standards, they prepare students for an uncertain future and ensure that they will be adaptable and successful in whichever path they choose after Kumsheen Secondary School.

Why publish a journal?

Student learning is something that should be celebrated by all. Student pride in their work increases dramatically when there is an audience, particularly when that audience includes family, friends, and community.

Specifically, the format of a scientific journal was used to introduce students to a new medium of communication – one that they might need to be fluent in should they pursue careers in science, or academia in general. Just as other media platforms have their own format, language, and conventions, so do scientific journals. Students are already fluent in communication via Twitter, Snapchat, Instagram, texting, and Facebook. The scientific journal is now one more communication medium at their disposal.

As an added bonus, students seeking academic careers may include their article in a "Published Works" section on a resume, which is sure to impress some employers.

Biological Aging Rates of Humans

Domonique Samson-Hayden

Abstract

Do all humans biologically age at the same rate? In my results I have found out that all humans don't biologically age at the same rate. There are many factors for why all of us humans don't age the same. There could be difference in sex, health, physical health, mental health, and emotions. Someone can be emotionally depressed and having lots of stress, causing that person to age faster, possibly getting wrinkles and grey hair. Another example is a person who works out daily, compared to someone who doesn't work out daily. All these factors can affect our biological aging rate. I was very interested to find out all of these reasons why humans don't age at the same rate.

Background

I can hypothesize many reasons why our biological aging rate is not at the same for all humans. My inquiry question is "Do all humans biologically age at the same rate?" Why is my question important? My question is very important because people want to know if we age at the same rate. It's important to me because I want to know why we all age differently or age the same. Aging is a big question to many. Why do we age? Why is aging a thing? Do we all age differently? All of these questions I asked myself. I believe we all age differently. If a person abuses their body, they will age faster. This will be apparent as their facial features will change, and their body will appear older. What's the difference between chronological age and biological age? The difference between the two is chronological age is how long the persons alive, and biological age is how old the person seems. I believe all humans biologically age at a different rate because no one can age at the same rate, many people don't realize abusing your body makes you age faster. I want to research and find a strong reason why we all age differently or age the same.

Methods

The method I used for my question was researching online to find others showing what they have found on this topic. I used search engines to find information on my subject. I primarily used the search engine Google Scholar to find my information on the human biologically aging process.

Results

The research I conducted will determine whether we all age differently or the same. Gender plays a huge part in our aging rate[3]. Female aging rates are lower then males for many reasons. Females tend to always keep their body heathy and always check in at hospitals or clinics if they don't feel okay. Men tend not to check themselves into a hospital or clinic if they don't feel okay. Another factor that plays a huge role in our aging process is that keeping a positive attitude can make your skin not wrinkle from frowning, also it lowers your stress level which prevents hair loss[1]. Drinking alcohol or abusing prescription drugs can increase your aging rate. Keeping a healthy diet also can help with your aging rate. Avoiding the consumption of processed food will reduce biological aging. Telomeres also play a huge part in your aging because if you start losing your telomeres you become more vulnerable to diseases and possibly even cancer[2]. Telomeres are the number one most important thing in our life, if we ruin out of telomeres there is a huge risk of dying because telomeres protect us from genetic diseases, sicknesses, and other dangers. Without telomeres, our life wouldn't last very long, people who lose their telomeres faster will die earlier then someone who kept their body healthy to maintain a healthy amount of telomeres. In the end of my research, I found many crazy reasons why humans age faster. Everyone should realize we don't restore telomeres, if we lose all of our telomeres then our lives are in danger.

Discussion

My question was "Do all humans biologically age at the same rate?" I was very interested in this question because I wanted to find out, if we do or not age at the same rate. In my research I came to a conclusion of my topic. I found out there are a lot of reasons why humans don't biologically age at the same rate. We all don't age at the same rate because it has to do with each persons health and physical health, they are all different a person can be a smoker and the other is a alcoholic. Humans don't biologically age at the same rate because of our difference in health.

References

[1] Garcia, O. Z. (n.d.). "Factors that affect the aging process". https://ozgarcia.com/factors-that-affect-the-aging-proc.

[2] Learn.Genetics (2012). "Are telomeres the key to aging and cancer". http://learn.genetics.utah.edu/content/basics/telomeres.

[3] Gur, R. (2002). "Gender differences in aging: cognition, emotions, and neuroimaging studies". https://www.ncbi.nlm.nih.gov/pmc/articles/PMC3181676/.

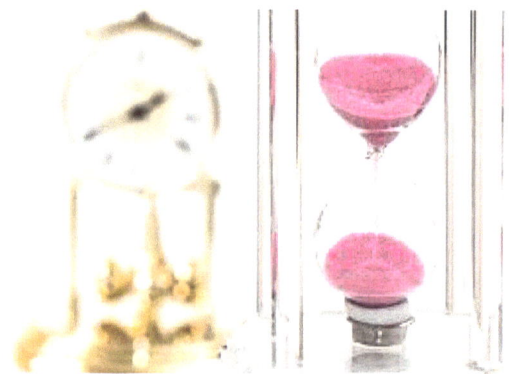

Creating GMO Humans

Ara Michell

Abstract

Is it possible to genetically modify a human? Yes, it is possible, but many religious groups and organizations strongly disapprove of it. There are still some experiments going on around GMO humans, but only under watchful eyes and the strict order to not pass the line between doing so to correct DNA mutations and wanting to give an embryo enhanced senses.

Background

Is it possible to genetically modify a human? We know that you can modify animals and food, but can we do it to humans to get rid of diseases and other potentially fatal deficiencies or deformities? The potential is there, scientists do it to mice all the time to test diseases and other such things. The only thing really stopping scientists from developing technology to do tests is the immorality of it. If you were to try to test out the possibility of night vision and you didn't get the genetic code right, you could blind them or cause them pain. Or if you were trying to modify an immune system to make it stronger, you could compromise it instead, or make it so the immune system attack normal tissues. But, if you were to take people with naturally excellent immune systems, you could examine the DNA that made it up and try to replicate it in an embryo.

With this, you could breed humans and thus create humans with great immune systems. But for this to happen you would also need to modify the DNA that affected aging so the human subjects grew at a faster rate than normal. This would make the breeding time shorter. But, then you have the problem of the human subjects growing too fast and dying of old age. For that to not happen, you'd need to extend the human life span. If you die of old age, it's because your cells have divided too much. Your cells can only replicate so many times. You would need to add more telomeres, long sequences of DNA that protect your chromosomes from damage. When your telomeres run out, your genetic information begins to be lost, thus damaging the health of the organism.

Methods

To research this topic, I needed a connection to the internet, google, and google scholar.

Results

An article I have found about genetically modifying a human is by Steve Connor on July 26th, 2016. In this article it talks about how researchers in Portland, Oregon used the CRISPR technique[1] to edit a one cell human embryo, called "germlining",[3] to correct DNA errors present in the father's sperm. The technique mentioned is a term loosely used for the various systems that can be programmed to target specific genetic code. This technique is the "Hallmark of a bacterial defense system that forms the basis for CRISPR-Cas9 genome editing technology."[5] To date, there have only been three previous recorded reports of embryo editing in China. However, none of these embryos stood a chance of being put into a womb and growing for more than a few days, as a range of religious groups, civil society groups, and biotech companies refused this.

In earlier Chinese reports, they found that using CRISPR had caused editing errors in the cells of the embryo. This caused many arguments about how germlining is a dangerous and unsafe way to create a human. James Clapper, U.S director of national intelligence, on February 9th, 2016, added gene editing to the list of "weapons of mass destruction and proliferation"[2] in the annual report of the world-wide threat assessment of the U.S intelligence community. "It's deliberate or unintentional misuse might lead to far-reaching economic and national security implications."

But, in another article "U.S. Panel Endorses Designer Babies to Avoid Serious Disease" by Antonio Regalado, the article states that "a panel of the National Academy of Sciences, […], recommended that germ-line modification of human beings be permitted in the future in certain narrow circumstances to prevent the birth of children with serious diseases"[4]. The panel had stated as well that "at this time" there would be a firm line between preventing disease and giving enhanced abilities that would not be crossed, and the scientists would be under strict supervision.

Discussion

With these few articles, the question "Can humans be genetically modified" has an answer. Yes, humans can be genetically modified. But, as it stands now, the only thing getting in the way of humans having greater intelligence is many morals and the potential of a human being genetically modified so much it becomes a danger to the world, with malicious intentions or not. Too many people repose the idea of genetically editing a human without the purpose of fixing errors in the DNA. For now the case stands as it is; we won't be seeing humans with accelerated healing or anything any time in the near future.

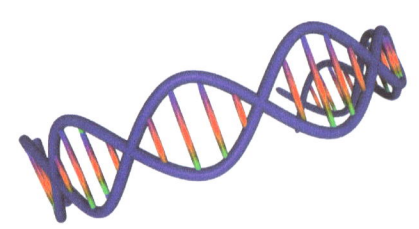

References and Acknowledgments

I would like to acknowledge my teacher, Mr. Houlton, for helping me out.

[1] Connor, S. (2017) First Human Embryos Edited in U.S. https://www.technologyreview.com/s/608350/first-human-embryos-edited-in-us/

[2] Regalado, A. (2016) Top U.S. Intelligence Official Calls Gene Editing a WMD Threat https://www.technologyreview.com/s/600774/top-us-intelligence-official-calls-gene-editing-a-wmd-threat/

[3] Regalado, A. (2015) Everything You Need to Know About CRISPR Gene Editing's Monster Year https://www.technologyreview.com/s/543941/everything-you-need-to-know-about-crispr-gene-editings-monster-year/

[4] Regalado, A. (2017) U.S. Panel Endorses Designer Babies to Avoid Serious Disease https://www.technologyreview.com/s/603633/us-panel-endorses-designer-babies-to-avoid-serious-disease/

[5] Institute, B. (2018) Questions And Answers About CRISPR https://www.broadinstitute.org/what-broad/areas-focus/project-spotlight/questions-and-answers-about-crispr

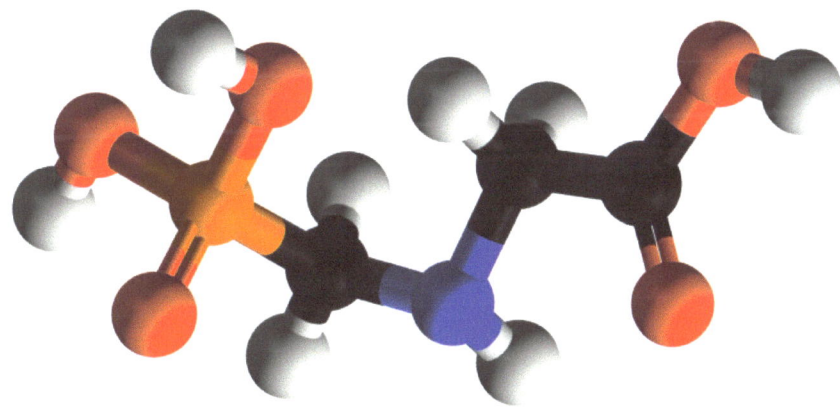

Junk DNA

Steven Arnouse

Abstract

My inquiry project is about Junk DNA and if the term "junk DNA" is a valid one. I have concluded that it is not a valid statement it was made decades ago and new research has been done on it proving that Junk DNA now renamed non-coding DNA is vital to cell genome for things like making proteins and it makes up 98% of your cell genome.

Background

My inquiry project is about if there "is such a thing as junk DNA". I do not have much prior knowledge on "junk DNA", in fact this is the first time I've ever heard about it; I thought it was an interesting topic. I hypothesize that junk DNA is the DNA inside a genome that help to keep it going and makes repairs if something goes wrong with the genome.

Methods

I researched my project "is there such a thing as junk DNA" using google scholar. There were quite a few different sites on junk DNA but I checked through some of those different sites and found that most of them didn't have much information on junk DNA. I found a few sites that did have research papers on junk DNA, but most you had to buy to read. Only two that I found are free. I couldn't understand everything that was talked about in some of the research papers I found so I thought I'd try just regular google and I found a few more sites that simplify what was said in the research papers.

Results

The term "junk DNA" refers to the regions of DNA that are non-coding. These regions help to produce RNA (ribonucleic acid) components such as transfer RNA, regulatory RNA, and ribosomal RNA. Junk DNA is also any part of the genome that doesn't code to make protein[1]. The term "junk DNA" was first used in the 1960s, but was formalized by Susumu Ohno in 1972 after she noticed that the amount of mutation occurring as a result of deleterious mutations set a limit for the amount of functional loci (fixed region in the genome) that could be expected when a normal mutation rate was considered. There are also other regions that don't produce RNA molecules and their function is unknown. Researchers have studied these areas for many years but what these regions do is still a mystery. Junk DNA makes up at least 90% of the genome according to ENCODE (Encyclopedia of DNA Elements). Where as in bacteria non-coding DNA makes up only 2% of the genome. My sources weren't the best; the big projects like ENCODE had to be bought and the other sites I found weren't as reliable but they had good information and it lined up with the other sites I was looking at.

Discussion

My research project examined if there is such a thing as junk DNA, and if so what that means. My research showed that junk DNA is actually non-coding DNA. Non-coding DNA is all of the DNA in a genome that doesn't make protein which is around 98%; only around 1.8% of DNA codes for a protein. After looking at all of the different papers available, I found out that what we call "junk DNA" isn't junk at all, it's a vital part of the genome. Junk DNA is like the map and the proteins like the drive, but the protein can't get to where it's going without the non-coding DNA. It was just called junk DNA 46 years ago because they didn't understand it was important. We still don't fully understand what parts of the non-coding DNA does, but we are learning that it isn't junk DNA. My research is a good starting point for somebody that doesn't know anything about junk DNA. My research could be improved if I had access to better sites and bigger scientific research projects.

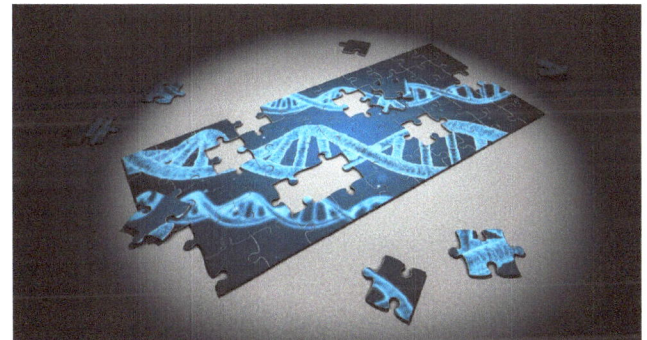

References and acknowledges

[1] Mandal, A. (2014). Functions of Junk DNA
https://www.news-medical.net/life-sciences/Functions-of-Junk-DNA.aspx

[2] Palazzo, A. (2014). The case for Junk DNA
http://journals.plos.org/plosgenetics/article?id=10.1371/journal.pgen.1004351

[3] Gregory, T. (2014). The case for Junk DNA
http://journals.plos.org/plosgenetics/article?id=10.1371/journal.pgen.1004351

Causes of Acne

Jacqueline Webster

Abstract

I have learned a lot doing my research about the topic of what causes acne. The results I have come across indicate that the more dairy products you consume, the more acne you get. The reason behind it is that dairy products have hormones in it which cause acne. Other factors which contribute to causing acne are your diet, medication, and your hormonal changes as a teenager. Acne forms due to a lack of taking care of your skin, due to the oils that clog up your pores and large glands that are attached to your hair follicles.

Background

I already knew a few facts about what causes acne. Acne is caused from the grease and the oils that you consume, particularly fast foods, other fatty and greasy foods, and more. Also, acne can come from the makeup you use since it clogs up your pores and creates more acne. Foundation makeup has a liquid consistency which forms clogs in our pores when we use them[1]. Acne is caused by the unhealthy foods you eat, such as candy, chocolates, and mainly the junk foods that you decide to consume containing high amounts of sugars. So far I know that women who have a regular menstrual cycle have a higher chance of getting more acne. I believe this happens because the pores are more vulnerable and are opened up.

Why is my question so important to know? It is important because everyone wants the knowledge to prevent acne, and of how acne affects you. People also want to have information on how acne is formed and have the opportunity to change their ways of consuming their foods to prevent acne. I believe that once people know more about what forms acne they would want to prevent it from happening. Once they get to know more about it, I think they'd be more happy with the results they get from it. The more healthy foods you eat and the less sugary drinks, the less acne is formed.

Methods

I collected information using articles available online through Google.

Results

I have done some research on some articles which said that an unclean diet and dirty skin are two of the main causes of acne[5]. Also, the lack of washing your face clogs up all the pores and causes more blackheads[1]. Regular hygiene will help reduce acne, the more water you consume will prevent acne forming. There are other factors that cause acne which if you are stressed it can lead to you having more acne than usual[3]. One other factor causing acne is if you regularly smoke cigarettes, since the nicotine creates pores in your skin[3]. This causes the acne to build up more glands and form more pores in your skin.

These are the results I have learned so far about the way acne can develop and who it affects more. From consuming dairy products such as milk, chocolate milk, ice cream can have an effect on your acne. This is from the chocolates and the fatty foods that cause the pores in your skin to develop more acne. Also, the choices in your diet can affect the way your skin produces oils. Acne is more common in females than males. Teenage acne is also more common and affects 90% of adolescents. Acne can affect all ages. One study I found on how to prevent acne found a link with dairy consumption. Researchers have found that reducing milk consumption in a high school had positive results for reducing acne. The causes of acne can be from an excess amount of stress, causing you to accumulate oils in your skin, which is connected to your hair follicles and causes glands. Acne can appear regularly on your forehead, chest, upper back and shoulders which are the areas of your skin in which oils build up most due to sweat[3]. Diets rich in skim milk and carbohydrate-rich foods may worsen the effects of acne.

Discussion

My question I have chosen for this project is "what causes acne?". I was interested in how this happens to more teenagers than adults. The reason people get acne and others don't is because of their hormonal changes as teenagers that are caused by androgens that increase during puberty. Also, teens have a higher risk of getting acne than adults, but in some cases if your parents have had acne there is a chance of being at higher risk yourself.

I have come to a conclusion that teens are most likely to get acne, having a greater hormonal change than adults and have less options to get rid of it. The other causes of acne can be from the medication you take, such as birth control. During pregnancy women are most likely to get acne which are affected by stopping birth control. There are two factors that mainly cause acne, one of which is that you inherited a predisposition for acne from your parents, and the other that you have oily skin which causes more large pores. I found that there are also factors such as consuming dairy products which affects both teens and adults the same[4]. I have found out that the causes of acne may decrease as you pass your teen years, but as you get older you have a less chance of getting rid of your acne as your hormonal rate changes. There are factors which worsen acne, such as hormones, medications, dietary, and in most cases stress[3]. Androgens are a type of hormone that increase the glands in teenagers both female and males, which also enlarge glands of sebum.

References

[1]Magin, P (2006) "The Causes of Acne" . https://search.proquest.com/openview/188960ee6f62aa4955bf05510ae4ad19/1?pq-origsite=gscholar&cbl=30763

[2]Beylot, C. (2002). "Mechanisms and causes of acne". http://europepmc.org/abstract/med/12053788

[3]Mayo, C. (n. d.) "Acne". https://www.mayoclinic.org/diseases-conditions/acne/symptoms-causes/syc-20368047

[4]Staff Writer. (n. d.). "Adult Acne VS. Teen Acne". https://www.acne.com/types-of-acne/adult-acne-vs-teen-acne/

[5]Aktins, E. (2017) "Milk sales continue to slide as diets, society shift away from dairy". https://www.theglobeandmail.com/report-on-business/milk-sales-continue-to-slide-as-diets-society-shift-away-from-dairy/article26117550/

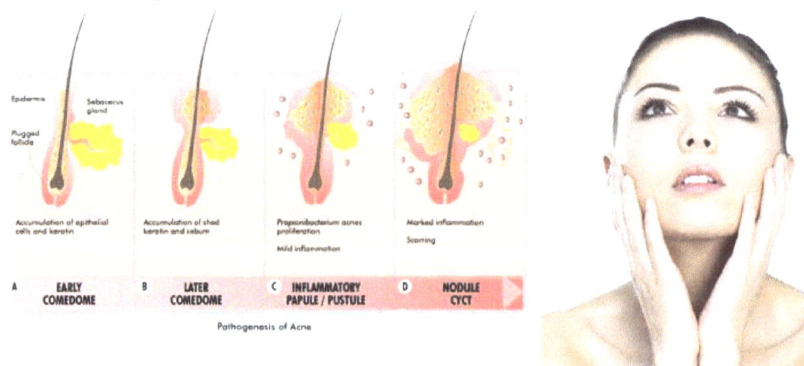

Benefits of Short-Term Fasting

Chase Johnny

Abstract

How does fasting for 24 hours affect your body? Depending on the type of help you get with the process of fasting, I think that a 24 hour fast is recommended. However, if you feel that you don't have the help you need, you should not fast for 24hours. If it is not done correctly, you are going to have some health problems.

Background

This inquiry question is "How does fasting for 24 hours affect your body?". This question is important because if a person wants to fast, then they must know what they are getting themselves into. From my knowledge of fasting, I know that when you are in the process of fasting, you cannot eat anything or drink anything except for water. If the process is not done right, there can be some health issues with the person who fasted. I personally don't know what kind of health issues people who fast are at risk for. Fasting can result in losing weight and cleaning your system. I think if someone were to fast for 24hours, that person will lose fat because they won't be eating anything for 24 hours and the body will eat away your stored fat until you put some new food in your system. If it is not done with a professional who knows the proper way to fast, things can go very wrong. I recommend getting a professional to help with the process before starting your fast.

Results/Methods

A 24 hour fast, also known as intermittent fasting, is becoming a popular weight loss method for people who have reached plateaus in their weight loss journeys[1]. What works best for clients is fasting on a day you are very busy; a day full of errands is great for fasting. Choosing a busy day to begin your fast keeps your mind off of food. Boredom is your enemy during fasting because your mind will try to convince you that you are hungrier than you are when it wants to eat out of boredom. The best time to start your fast is after dinner. By beginning a fast after dinner, you can go to sleep with a full belly and wake up and do your errands. There are many benefits according to NPR's article, such as weight loss, enhancing your resistance to stress, helping to prevent cancer, improving your brain function, improving your immune system, increasing your life span, raising growth hormone levels in the body, and maintaining lean muscle tissue.

Beginners should choose a day that is stress free, with relatively few activities planned[2]. Do not consider heavy work or exercise travel during that day. Light activities, such as reading, slow yoga movements, working on your computer, going on walks, meditation, watching T.V, driving short distances, etc, are acceptable, but avoid stressful activities and try not to consume a lot of calories. A human body can go for weeks without eating anything, therefore a one-day fast should be fairly easy to accomplish. Once you have started, you will experience great health benefits. Under the supervision of an expert, one with medical conditions may also fast. If you have a serious illness, you should consult a physician first before attempting to fast. In a one-day fast, you must only drink water for 24 hours. No solid foods, nor drinking juice or milk. Pure water is the best. You can also boil water. Do not add anything to the water while it is boiling. One may drink 1 ½ liters to 3 liters of plain water in one full day, according to one's capacity. There is no harm in drinking even more, if needed.

Note that the information from source two contradicts source one. Source two recommends a stress free day when beginning a fast, whereas source one recommended a busy day. In comparing the two articles, I felt that source two lacked credibility and source one should be considered more reliable.

The best time to start the fast is during early morning[3]. After getting up from bed and brushing your teeth, drink about 2 cups of water. This will help with bowel movements. Throughout the rest of the day, you can drink water anytime you want/need. There are no restrictions on the amount of water to drink, or the number of times to consume water. Continue this for a whole day until the next morning. During the process of 24 hour fasting, you will notice a few reactions in the body, which is normal. These reactions are in response to your body's habit of eating food three times a day. When there is lack of food consumption, you may feel weakness, dizziness, nausea, etc. A drop in blood pressure may also happen and slight headaches can occur. Do not be scared of these reactions at all, it is normal for these reactions to happen. People who fast on a regular basis will not feel these symptoms, however beginners should just lie down and rest when the symptoms get intense. During the fasting period, there is hardly any digestive activity. You should be gentle on the stomach when breaking the fast. Do not eat a lot. The best way to break a fast is with some sort of citrus juice. Half a spoon of honey may also be added to the citrus juice. If needed, even fruits can be taken, seeing as it is easy to digest. Even boiled vegetables are fine, try not to use any spices, and if you do use spices don't use a lot.

Discussion

The question everyone wants an answer to "How does 24 hour fasting affect your body?". According to the articles I have read, I think it is safe to fast, as long as you have talked to professionals and doctors about it first and discussed the procedures that you should be following. There are some things that can happen though if you do not do it right. Extended fasting can lead to electrolyte imbalance, thinning hair, as well as the downy hair associated with eating disorders[3]. Prolonged fasting can also result in cardiac arrhythmia, renal failure, and starvation leading to death. Fasting while taking medications can cause dangerous complications. Ultimately, the decision whether to fast or not is personal, and practitioners do so at their own risk.

References

[1] McCanus, L. (2017). "24-Hour Fasting for Weight Loss – Is It Safe?". https://avocadu.com/%E2%80%8B24-hour-fasting-for-weight-loss/

[2] Unknown, (n.d) "One Day Water Fast and its Benefits". http://www.yogicwayoflife.com/one-day-water-fast-and-its-benefits/

[3] Baker, L. (2014) "The Ultimate Cleanse? The Pros and Cons of Fasting, Plus a How-To Guide". http://www.onegreenplanet.org/natural-health/the-ultimate-cleanse-the-pros-and-cons-of-fasting-plus-a-how-to-guide/

Genetic Basis of Races

Patashi Pimms

Abstract

Many studies have been done to attempt to prove whether or not there is a genetic basis to race. Arguments have been made on whether race and ancestry are the same concept and whether those concepts are important to biomedical research. This article provides facts about both arguments and all points of view. Is race just a social concept or is more important than that? Many studies have been done to suggest that there is a genetic basis to race, however nothing has been definitively proven or disproven.

Background

There are many different ideas and definitions of race. Two attempts to define race include "a group of people identified as distinct from other groups because of supposed physical or genetic traits shared by the group" and "a socially constructed category of identification based on physical characteristics, ancestry, historical affiliation, or shared culture". Race is commonly seen as just a social construct and nothing more. However, I hypothesize that it is more. People within the same race share similar physical and even biological traits. For example, First Nations people commonly have brown eyes, skin and dark hair. There have been studies that have shown that a possible reason for high diabetes rates within First Nations populations are due to their bodies never having been required to dispose of sugars in past generations, therefore they are having trouble to now. Finding a genetic basis with races could be the start of individual treatments of people with diseases and disorders. I believe that research will prove that there is a genetic basis within races.

Methods

I collected research from the search engine "Google Scholar".

Results

Biomedical scientists have contradicting opinions about race. A percentage of scientists believe that race is "biologically meaningless"[1], whereas some believe and advocate the use of it in medical research. I am confident that all my resources came from a credible source. All sources came to the same conclusion that race has no genetic basis, but ancestry does except for Lynn Jorde and Stephen Wooding[1] who stated that they are basically the same thing.

Studies have shown that 85-90% of genetic variation is found within Africa, Asia and Europe and only 10-15% of variation is found between them[1]. Patterns have shown three important trends. The first trend shows that populations cluster due to their geographic distance from one another. The second trend shows that African populations have the most diversity within genetics. Finally, the third trend shows that the largest genetic distance is between those who are of African descent and those who are not.

There is a troublesome concept of race and whether or not scientists can use it to categorize humans[2]. Though many believe that there is a basis, no one has been able to prove it. The concept of race has not proven any value in public health, but more of a social construct. In biological and social sciences they prefer using ancestry, rather than race. At the nucleotide sequence level, all humans are ~ 99.6-99.8% identical[3]. A scientist named Blumenbach categorized people into 'five races' based on phenotypic physical features.[3] Morphology is not the best way to categorize humans, considering most physical attributes result from adaptations from their environment.[3]

Discussion

As the research shows, not everyone has the same opinion and definition of race. From my findings, I have concluded that scientists see race as just a social concept, but I don't see race as just that. The traditional definition of race was to group people in the same geographical area that share the same culture. I do understand that since this traditional definition was defined the world has changed, people don't always live where their genes may have originated from. Ancestry is more commonly used because it is seen a "more accurate", although race is defined as people who have the same ancestry. I believe the argument on whether race and ancestry are the same is irrelevant to my inquiry question, from here on I refer to the both as the same concept.

On a geographic point of view 85-90% of all genetic variation is found within three continents. The largest genetic variation is found between those of African and non-African descent. These facts prove that people of different race or ancestry have a variation in genetics. From a medical standpoint, it is known that people of certain ancestral backgrounds may be more susceptible to certain diseases[2]. An example of this is sickle cell anemia. Although this condition affects people of all ethnicities, this disease is more common in certain ethnic groups. However, stating that there is a basis to race based on morphology is not accurate, because physical attributes are caused by the environment. There have been many studies that try to prove whether or not there is a basis, but the bottom line is that no one really knows for sure yet.

References

[1] Jorde, L. and Wooding, S. (2004). Genetic variation, classification and 'race' *Nature Genetics **volume36**, pagesS28–S33*

[2] Cooper, R. (2003). Race, genes, and health—new wine in old bottles? *International Journal of Epidemiology, Volume 32, Issue 1*

[3] Tishkoff, S. and Kidd, K. (2004). Implication of biogeography of human populations for 'race' and medicine. *Nature genetics.*

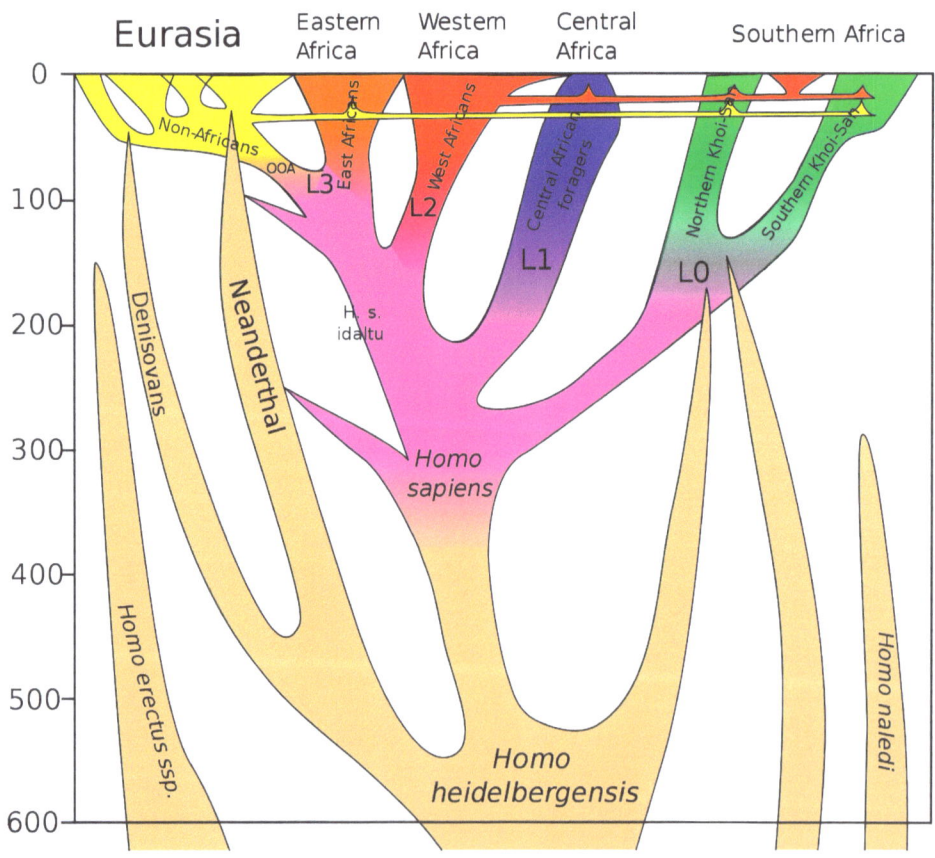

Fetal Alcohol Exposure

Brianna Smith

Abstract

Alcohol consumption during pregnancy may result in severe prenatal growth deficiency. In this study, prenatal exposure to alcohol is found to be one of the preventable causes of birth defects. A cluster of birth defects resulting from prenatal alcohol exposure was recognized as fetal alcohol syndrome (FAS). Drinking alcohol during pregnancy can cause miscarriage, stillbirth, and a range of lifelong physical, behavioral, and intellectual disabilities.

Background

Alcohol consumption during pregnancy causes Fetal Alcohol Syndrome (FAS) and Fetal Alcohol Spectrum Disorder (FASD). Many people know that you're not supposed to drink during pregnancy because of the possible birth defects, both physical and behavioral. I know alcohol consumption causes many problems but how does alcohol consumption during pregnancy cause birth defects? This is important because there are many factors to consider when you're pregnant. It should be everyone's goal to reduce the incidence of birth defects due to fetal alcohol exposure.

Methods

I found all the information I know about alcohol consumption during pregnancy on the internet.

Results

Alcohol in the mother's blood passes to the baby through the umbilical cord[1]. Drinking alcohol during pregnancy can cause miscarriage, still birth, and a range of life long physical, behavioral, and intellectual disabilities. Children with FASD might have abnormal facial features such as a smooth ridge between the nose and upper lip, small head size, shorter than average height, low body weight, poor coordination, hyper active behavior, difficulty with attention, poor memory, difficulty in school (especially in math), learning disabilities, speech and language delays, intellectual disability or low IQ, poor reasoning and judgement skills, sleep and sucking problems as a baby, vision or hearing problems, problems with the heart kidney or bones.

There is no known safe amount of alcohol to drink while pregnant or trying to get pregnant. Alcohol consumption during pregnancy can also cause seizures and other neurologic problems, delayed development (kids may not reach milestones at expected time)[3]. Alcohol consumption during the first trimester can cause major birth defects, later in the pregnancy drinking alcohol can cause poor growth and brain damage that could lead to learning and behavioral problems[2].

A child who is thought to have FASD's may be referred to a development pediatrician, genetic specialist, or another specialist who can help identify the problem and confirm a diagnosis. There is no cure for FAS or FASD's but many things can be done to help a child reach his or her full potential, children can benefit from services and therapies such as speech language, occupational and physical therapy.

I researched my inquiry question on many of websites that were all almost the same but there was some websites that said the alcohol passes through two different places in the body. One website says that the alcohol easily passes through the umbilical cord and another website says that the alcohol passes through the blood and veins. I concluded that the alcohol passes through the umbilical cord not the veins.

Discussion

My research question is "how does alcohol consumption during pregnancy cause birth defects?" when you drink alcohol so does your baby, because babies are small compared to adults that means alcohol remains in a baby's blood much longer than in the blood of its mother. Alcohol consumption during pregnancy causes birth defects, alcohol passes through the umbilical cord even if you drank before you knew you were pregnant. There are some birth defects that are more severe than other birth defects, it all depends on how much alcohol you consume during pregnancy. All of my findings came from a bunch of different websites, a sentence or two from each website. Alcohol consumption during pregnancy is defiantly not worth it, your child deserves a healthy long life with no problems with their body.

References

[1] Center for Disease Control. (2018). "Fetal Alcohol Syndrome Disorders". https://www.cdc.gov/ncbddd/fasd/alcohol-use.html#

[2] Gavin, MD. (2016). "Fetal Alcohol Syndrome". http://kidshealth.org/en/parents/fas.html

[3] Government of Canada. (2012). "Alcohol and Pregnancy: The sensible guide to a healthy pregnancy" https://www.canada.ca/en/public-health/services/health-promotion/healthy-pregnancy/healthy-pregnancy-guide/alcohol-pregnancy.html

Effect of GMOs on Human Cell Biology

A brief introduction to the safety of Genetically Modified Organisms

Mason Jory

Abstract

The question which drove this study was "What Effect Does the Intake of Genetically Modified Organisms Have on Human Cells?" The human and animal digestive tract rapidly degrades modified DNA as it would with a naturally occurring substance. There is evidence that traces of consumed GMOs can be found in host tissue, but this is much alike the DNA fragments found in non-GM plant material that can be detected in the same tissues. The notable harm that is caused by Genetically Modified Organisms arises from the chemicals that are applied to them, which can harm humans but not the targeted plant, as it has modified resistance. Therefore, Genetically Modified Organisms cannot be deemed dangerous alone, as they require the absorption of a pesticide or other harmful chemical to cause noticeable damage to cells.

Background

I predict that the presence of GMOs can have an effect on human cells, as they are not naturally occurring organisms, but I cannot justly assume if this effect will be negative or even noticeable.

Genetically Modified Organisms are any organism which have had their genomes altered by any means of Genetic Engineering. These organisms have had genes inserted, substituted or deleted from their DNA in a process that is not naturally achieved. This practice is widely accepted in pharmacy, agriculture and food industries. In the field of pharmacy, genetic engineering can provide less costly alternatives for manufacturing pharmaceuticals such as the Hepatitis B vaccine found in GM baker's yeast. Genetic modification can be used to eradicate traits of foods that are unfavorable, such as the wilting of lettuce would extend the shelf life of the vegetable.

Results

Research by market watchers indicate that about 70% of all processed foods found in a local grocery store contain GMOs[1]. There is no exact estimate of this number as food companies are not at liberty to disclose any GMOs within the product. The importance of this question is that Genetically Modified food is increasingly popular, and could very well be the face of food and agriculture. Therefore, all research contributing to the effects of GMOs of the human body should be deemed important, as Genetically Modified food is accumulating in our bodies more than ever before.

The popularity in GM foods is based on the convenience of pesticide application and the increased amount of desired traits in the food. However, pesticides applied to plants that do not kill them, are absorbed into the plant's cells. Researchers have observed damages to membranes of human cells, as the residue causes an inhibition of cell respiration as the result of being exposed to the pesticides[1]. A study connected soy consumption and a pesticide called "Roundup" to digestive disorders such as irritable bowel syndrome as well as fatigue, headaches, skin disorders and nausea[1]. Additionally, allergies can be formed from the result of consuming modified proteins, as they do not occur naturally in food and our bodies are not equipped to break them down.

Scientists can agree that genes are not specifically among the individual members of a species, but also shared among members of different species[2]. Fragments of DNA have the possibility to end up in animal tissues and milk products people consume[2]. Small fragments of Genetically Modified DNA are occasionally found in tissue when humans or animals digest genetically modified foods. The transferred genes proved to not be harmful, but these are genes that are not naturally created by humans. The digestion in both animals and humans is known to rapidly degrade modified DNA, and the uptake of DNA fragments from the intestinal tract into the body is a normal physiological process that occurs with all food[2].

There are animal studies which support the idea that small amounts of nucleic acid can pass through the bloodstream and even enter the tissue. For example foreign DNA fragments were detected in the digestive tract and leukocytes of rainbow trout fed by genetically modified soybean, and other studies report similar results in goats, pigs and mice[3]. Based on the analysis of over 1000 human samples from four independent studies, evidence was found that Genetically Modified DNA fragments which are large enough to carry complete genes can avoid degradation and enter the human circulation system[3]. In one of the blood samples the relative concentration of plant DNA is higher than the human DNA[3]. However, a much larger human based study would be needed for convincing results.

Discussion

Although the popularity of Genetically Modified foods has grown exponentially, research relative to the effects of these organisms in regard to human cell biology is uncommon. Animal research has proved that genes from these organisms can be transferred to a host via GM animal feed and that evidence of this can be found in the tissue. Despite how readily genes are shared and their ability to enter the human bloodstream, Genetically Modified Organisms have not proved to negatively impact human cells on their own, but the pesticides associated with them can be easily absorbed into the body causing membrane damage in cells and oxygen inhibition. There are potential health effects that could result from the insertion of a foreign gene into an organism, but these health effects have not been studied to the extent that would prove them to be harmful. Genetically Modified Organisms have the potential to carry pesticides that harm humans, but the modified genes themselves have not caused any notable harm to human cells.

References

[1] (2016) G. DAHOM "Genetically Modified (GM) Foods Can Cause Cell Damage". http://totalhealthmagazine.com/GMOs/Genetically-Modified-GM-Foods-Can-Cause-Cell-Damage.html

[2] (2013) Food Standards Agency "GM Material in Animal Feed". https://www.food.gov.uk/safety-hygiene/genetically-modified-foods

[3] (2013) S. Spisák Public Library of Science "Complete Genes May Pass from Food to Human Blood" http://journals.plos.org/plosone/article?id=10.1371/journal.pone.0069805

Dr. Gloria is a Certified Dental Professional, Doctor of Natural Health, Homeopath, and Certified Dietary Supplement Counselor. She is an acclaimed, syndicated talk show host, Dr. Gloria—Health Detective, author of 18 books, 8 courses and over 1,700 health articles.

The Food Standards Agency is a government based research institution which has high credibility in the science community. There are the independent committees, working groups and forums that advise the Food Standards Agency and help ensure that the Agency's advice to consumers is always based on the best and most recent evidence.

The Public Library of Science was created by Nobel Prize winner Harold Varmus. All submissions go through a pre-publication review by a member of the board of academic editors, who can elect to seek an opinion from an external reviewer. It was included in the journal citation reports in 2010.

Iron Deficiency

Nathanial Brown

Abstract

Why does low iron make you so tired? This was my scientific question. I feel that this question can relate to a lot of people who have iron deficiency. The examples that I put into this script are facts to how you can get iron deficiency, how you can negate iron deficiency, and how you know that you have iron deficiency. Expanding on the idea of low iron, there are other things that will also be affected if you have low iron Example of these things would be your hemoglobin. What hemoglobin does for your body is it circulates the oxygen your body needs. This affect is called anemia.

Background

My question is why does having low iron make you so tired? To summarize what I already know of low iron, women most likely have it more than men do, and the proper term for low iron is "Iron Deficiency". My extent of knowledge on iron deficiency is that the lack of iron makes you have less red blood cells which end up having less blood oxygen flowing through your body. Linking my knowledge between my inquiry question and my prior knowledge is that it seems to be that they seem to be pretty linked since I've already done research around how low iron makes you tired. I think my question is important because low iron can cause a real problem to most people if they don't know what's causing them to have such low iron.

My hypothesis on what low iron is that it would cause you to have less sleep and have it so that you have barely any energy to get through the day. A way to get your iron back is through tablets in which you can swallow or put in a cup of water.

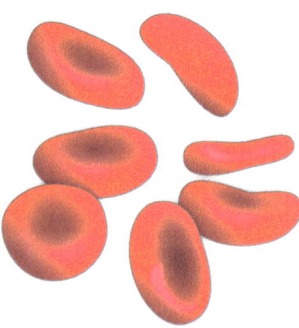

Methods

I found this information by going to google scholar and being able to find the first site at Mayo clinic, Then to how I found the other two sites was by doing a google search and finding trust worthy sites that share in a comparison with one another.

Results

What the Mayo Clinic[1] has to say about Iron deficiency is that when you have low iron, the blood lacks adequate healthy blood cells. Then the symptoms for what iron deficiency does is that it causes weakness, extreme fatigue, pale skin, and a whole list of other symptoms. Iron deficiency is caused by blood loss, the lack of iron in your diet, and the inability to absorb iron.

Dr. OZ[2] says that iron deficiency is a reason why you have so little red blood cells pumping because of how little iron that you have in your body system. The symptoms that are listed on this site are more focused on the women's side. These symptoms include having heavy periods, which would relate to major iron deficiency. The other side effects would be not being able to focus, having paler skin, and always feeling cold. The site also refers to what food would prevent iron deficiency. Examples of recommended foods are red bell peppers, tomatoes, and broccoli.

The Atlantic article[3] says that you should not be so hasty with taking the iron supplements. If you do take the iron supplements, you should average around 80 milligrams of iron per day. Continue this cycle for 12 weeks to see the results. They were able to find this data by having a total number of 200 women take the test and assessed the number of women ages 18 to 50 still complaining about their fatigue.

I find that these sites are pretty reliable because of how much similarities each of these articles have such as, them all talking about hemoglobin which refers to the lack of red blood cells. Yes they may have different areas of focus such as Dr. Oz and Atlantic Articles focusing on women while Mayo Clinic talks about an overview into solving the basic procedure and symptoms that you may have.

Discussion

How does low iron make you so tired? The answer that I came up with from my results is that when you have low iron it results in less red bloods cells forming in your body. These cells are responsible for carrying the oxygen that flows through your body. The symptoms that it would cause to your body is that you won't get as much sleep. What the other sites say is that low iron can effect most people and the people who don't treat there iron deficiency will have symptoms such as, dizziness, headaches, and pale skin.

References

[1] Mayo Clinic Staff (2016) Iron Deficiency Anemia
https://www.mayoclinic.org/diseases-conditions/iron-deficiency-anemia/symptoms-causes/syc-20355034

[2] Dr. OZ (2012) Fight Fatigue: Reverse Your Iron Deficiency
http://www.doctoroz.com/article/reverse-your-iron-deficiency

[3] Fontenot B. (2012) Iron Loss Explains Why You're Tired All the Time
https://www.theatlantic.com/health/archive/2012/07/iron-loss-explains-why-youre-tired-all-the-time/259616/

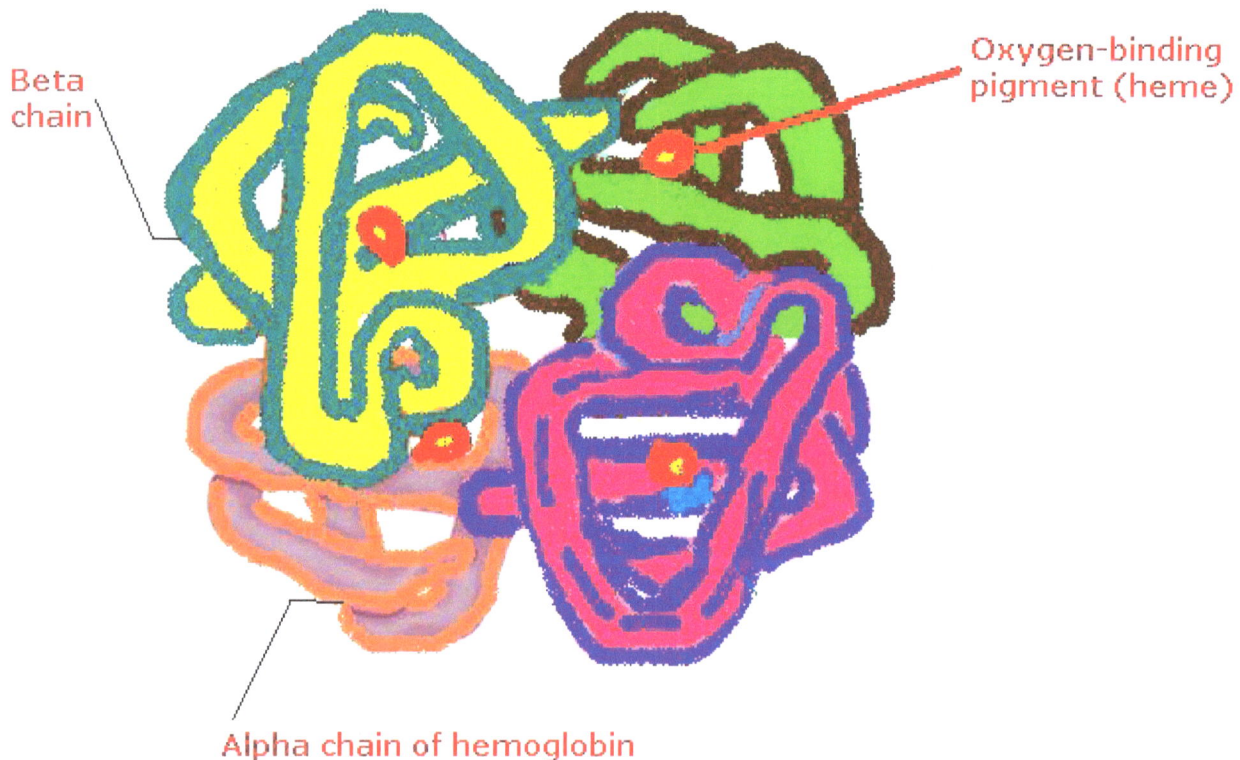

Human Cloning

McKenna Adams-James

Abstract

Is it possible to clone human beings? Scientists have cloned so many non-human animals, therefore I hypothesized that they might also be able to clone our species. I concluded, however, that we are not yet able to clone grown humans because we don't have the technology.

Background

Cloning is the process of producing a genetically identical copy of something or someone. Scientist have only been cloning non-human species. Scientists have cloned 20 species[1], including Dolly the sheep (which was successful), a family of pigs (Millie, Alexis, Christa, Dotcom, and Carrel), as well as Mira the goat.

I have heard of cloning during science classes, in science fiction books, and in movies such as "Jurassic Park" and "Jurassic World" when they did attempt to clone one of the dinosaurs, but failed. Cloning seems so unreal and fascinating as if it should only exist in a parallel universe or in books and movies. Based on my what I already know about cloning, I hypothesize that we may possibly be able to clone humans.

Results

The Roslin Institute in Midlothian, Scotland cloned Dolly[3]. Dolly was cloned from a female sheep that was six years old. Dolly the sheep lived for six years; she developed a lung disease and sadly had to be put down. The technique they used to clone the sheep was through somatic cell nuclear transfer[4]. If they can clone a sheep why can't they clone a human?

The reason why scientists want to clone human embryos is because it would be a proficient way of manufacturing embryonic stem cells[3]. That's exactly what Shoukhrat Mitalipov of the Oregon Health & Science University did. They gathered and asked women if they were able to provide their eggs for the research. Once they found women who were keen to give up their eggs, the scientists distributed most of the data from the eggs, they then continued to replace the genetic material with other people's DNA from skin cells. They discovered a way stimulate the eggs (so it would turn into an embryo), but without the need to fertilize it with sperm. The stimulation was performed using a variety of substances and electrical pulses. Unexpectedly, the best stimulant turned out to be caffeine. The work that they did had received criticism because of the ethical concerns.

Natural clones/cloning, commonly referred to as "identical twins", is just a split fertilized egg, producing usually two (sometimes more) embryos that carry similar and identical DNA[2]. They have the equivalent genes as each other, but of course they are dissimilar from both parents.

Discussion

Based off of all my research I have done, it is not yet possible to successfully clone a fully formed human. I searched so many things but couldn't find anything on cloning an actual human. The closest thing to cloning fully developed humans is the cloning of human embryos. They have cloned so many species, and have recently cloned monkeys, but not humans. Based on the successful cloning of humans to the embryo stage, scientists are now working on cloning human babies[5]. Once we get the right technology, it might be possible. However, we will not be cloning any grown human any time soon.

References

[1] Knufken, D. (2009). "20 Animals That Have Been Cloned". http://www.businesspundit.com/20-animals-that-have-been-cloned/

[2] National Human Genome Institute. (2017). "Cloning". https://www.genome.gov/25020028/cloning-fact-sheet/

[3] Stein, R. and Doucleff, M. (2015). "Scientists Clone Human Embryos To Make Stem Cells". https://www.npr.org/sections/health-shots/2013/05/15/183916891/scientists-clone-human-embryos-to-make-stem-cells

[4] Science Daily Staff Writer. "Dolly the Sheep". https://www.sciencedaily.com/terms/dolly_the_sheep.htm

[5] Baker, M. (2014). "Stem Cells Made by Cloning Adult Humans". https://www.nature.com/news/stem-cells-made-by-cloning-adult-humans-1.15107

Function of Taurine in Energy Drinks

Sadie Drynock

Abstract

My cell biology inquiry project explores the question "what is taurine, and why is it used in energy drinks?". I didn't exactly know what was the purpose of taurine was in energy drinks but I hypothesized that it was used to provide people with energy. I concluded that taurine is an amino acid which supports and helps regulate the level of water and minerals in the blood.

Background

What is taurine, and why is it used in energy drinks? I first heard about taurine a few years ago when I began to use energy drinks. I didn't know anything about taurine other than it was advertised as an ingredient in energy drinks. I decided to research taurine and its role in energy drinks because I wanted to be more informed about what I was putting into my body. I hypothesized that taurine would have negative health impacts because I had been told that energy drinks in general are bad for your health.

Results

Taurine is an amino acid that supports neurological development and helps regulate the level of water and minerals in the blood.[1] Some studies suggest that taurine supplementation may improve athletic performance, which is probably why it's in energy drinks. Taurine is found naturally in meat, fish, and breast milk. Taurine is commonly available as a dietary supplement. In addition to taurine, it is also important to remember that there may be high amounts of caffeine or sugar in energy drinks. Too much caffeine can increase your heart rate and blood pressure, interrupt your sleep, and cause nervousness and irritability.

The role of taurine helps in the formation of bile salts, which are needed in the process of digestion.[3] Taurine also regulates antioxidant function, boosting the immune system. Taurine is used by the body to help support the function of the central nervous system. Taurine may be helpful for people with diabetes because it helps control blood sugar levels. Some researchers have shown that most people with diabetes have low levels of taurine. And therefore conclude that this could be one of the causes of the disease.

There are safety concerns regarding the consumption of energy drinks. There have been a number of deaths due to the intake of too much caffeine in relation to energy drinks, however there were no negative affects when energy drinks are consumed in recommended dosages by individuals. People with kidney problems may potentially have problems with amino acid supplementation from energy drinks.

Taurine is used by the body for cardiovascular function and to help with nerve signals such as pain.[2] It also helps support liver and bile function. It's one of the main amino acids that bind and neutralizes toxins. Your body normally synthesizes taurine. According to New York University Langone Medicine center, you can take up to three grams of taurine per day safely.

Discussion

Taurine is an amino acid that supports and regulates the level of water and minerals in the blood. It also helps with digestion through the formation of bile salts. Taurine regulates antioxidant function and improves function of the immune system. Taurine may be helpful with people with diabetes because it helps control blood sugar levels. There's safety concerns about the consumption of energy drinks due to excess amount of caffeine. People with kidney problems may potentially have problems with processing excess amino acids such as taurine.

References

[1] Zeratsky, K. (2018). "Taurine is an ingredient in many energy drinks, is taurine safe?" https://www.mayoclinic.org/healthy-lifestyle/nutrition-and-healthy-eating/expert-answers/taurine/faq-20058177

[2] Jacobs, J. (2017). "Taurine in Monster Energy Drinks". https://www.livestrong.com/article/416765-taurine-in-monster-energy-drinks/

[3] Unknown Author. (n.d.). "Why Is Taurine Used In Energy Drinks". https://www.findatopdoc.com/Diet-and-Nutrition/taurine-in-energy-drinks

Importance of Vitamin C

Farah Abbott

Abstract

Why is vitamin C important? This paper is going to be about why vitamin C is important. Vitamin C is important because if you don't get vitamin C you can start having health issues like nausea and anemia.

Background

I picked the question "why is vitamin C important?" because I thought it would be interesting. I know that we need vitamin C to be healthy. I know that you can get vitamin C from citrus fruits like oranges and kiwis. This topic is important because vitamin C is in a lot of the food that we eat.

Results

Vitamin C is also called an ascorbic acid. This vitamin helps prevent heart disease and cancer. Vitamin C is found in fruits and vegetables, and therefore it is easily to get in your diet. This vitamin helps your blood vessels, skin, and bones stay healthy. If you want to be healthy then you should at least have 200-250 mg of vitamin C per day. One of the source I found said that more than 2,000 mg of vitamin C per day can cause nausea[1].

People can take vitamin C to help their growth. It also helps with healing wounds and repairing teeth, cartilage, bones, asthma, and protects you from getting a cold. If you don't get enough vitamin C then it can cause bleeding gums, easy bruising, nosebleeds, and anemia. The easiest way to get your daily dosage of vitamin C is to go on a balanced diet. People who smoke require more daily vitamin C than people who don't smoke. Women who are pregnant also need more vitamin C. in their diet

Vitamin C is a water soluble vitamin. Water soluble means that it is not stored in your body, so you have to get it yourself through your diet. You can get vitamin C by eating foods or you can buy tablets of vitamin C. You can also buy it in a liquid or powdered form. Vitamin is an antioxidant, which is a type of molecule that helps protect DNA.

Discussion

Why is vitamin C important? It is important because people need vitamins in their system in order to stay healthy. If we don't get enough vitamin C then we can get sick and have symptoms like nausea and anemia.

References

[1] Weil, A., and Weber, B. (2012). "Vitamin C Benefits". https://www.drweil.com/vitamins-supplements-herbs/vitamins/vitamin-c-benefits/

[2] Wax, E. (2017). "Vitamin C". https://medlineplus.gov/ency/article/002404.htm

TABLE 7.1   Vitamin C Content of Selected Foods, in Milligrams per 3 ½-oz. (100-g.) Serving

Food	mg	Food	mg	Food	mg
Acerola	1300	Strawberries	59	Okra	31
Peppers, red chili	369	Papayas	56	Tangerines	31
Guavas	242	Spinach	51	New Zealand spinach	30
Peppers, red sweet	190	Oranges & juice	50	Oysters	30
Kale leaves	186	Cabbage	47	Lima beans, young	28
Parsley	172	Lemon juice	46	Black-eyed peas	29
Collard leaves	152	Grapefruit & juice	38	Soybeans	29
Turnip greens	139	Elderberries	36	Green peas	27
Peppers, green sweet	128	Liver, calf	36	Radishes	26
Broccoli	113	Turnips	36	Raspberries	25
Brussels sprouts	102	Mangoes	35	Chinese cabbage	25
Mustard greens	97	Asparagus	33	Yellow summer squash	25
Watercress	79	Cantaloupe	33	Loganberries	24
Cauliflower	78	Swiss chard	32	Honeydew melons	23
Persimmons	66	Green onions	32	Tomatoes	23
Cabbage, red	61	Liver, beef	31	Liver, pork	23

Source: U.S.D.A., Nutritive Value of American Foods in Common Units, Agriculture Handbook No. 456.

Alkaline and Acidic Diets

Brianne J. Duncan

Abstract

Which diet is better for you, alkaline or acidic? Within my research I found that though the effects between Alkaline and Acidic diets on your blood pH directly is non-existent due to the body needing to maintain a constant blood pH for your survival, there are many benefits to living with an alkaline-based diet. An alkaline-forming diet is said to provide numerous health benefits, immunity, and longevity of life, while acid-forming diets are high in PRAL (Potential Renal Acid Load) which can lead to "Metabolic Acidosis" and is linked to osteoporosis, risk of muscle wasting, and kidney stone formation. It is important for this subject to be researched further and become known to the public because a misunderstanding of the two diets and can lead to detrimental results, while the proper information about these diets can lead to increased personal health.

Background

There is one thing in particular that will always come up in conversations due to its importance in personal health. The topic of which diet is best for a healthy lifestyle with positive improvements, weight loss, and prevention of chronic illnesses. Two options that address these concerns are alkaline and acidic diets. So is an alkaline or acidic diet better for you? How does your body react to both styles of diet?

The general misconception of everyday people is that you can affect your blood pH levels by eating certain foods to change and in turn strengthen your overall health and ability to overcome illnesses. The human blood is strictly regulated at pH 7.35 to 7.45. It might be easy to believe that the alkalinity of foods can change your blood pH because levels above 7.45 are typically considered as alkalosis and levels below as acidosis. If your body's blood pH were to drastically change it could be life-threatening, so the reality is you are not changing the blood pH yourself but simply changing the amount that your body has to compensate for in order to maintain its standard blood pH.

The theory behind the ever growing topic of the alkaline diet is to ingest foods that are alkaline based as opposed to acidic based before going into the body. Examples of acceptable alkaline foods are fruits and vegetables, while acidic foods would be meat, dairy, chocolate, etc. You can affect the general pH throughout your body with the types of food you ingest, which can lead to either benefits or detriments to your health. It is important to clarify what it means to have these diets and the effects of each as there are many biased reports that do not give all or enough information to the topic. I believe that although the idea that the alkaline diet will affect your blood pH directly is incorrect there will be certain overall benefits in terms of personal health, and longevity.

Methods

The majority of sources I used were found on the PubMed website or simply through the Google search engine.

Results

The diets both have to do with whether the foods ingested by a person are alkaline-forming or acid-forming. This refers to the effect each food has on the body once it has been consumed. Typical acid-forming diets that are high in PRAL (Potential Renal Acid Load) include: animal protein, dairy foods, and refined grains. Over time this acid-forming diet can lead to chronic, low-grade metabolic acidosis. [1] Metabolic acidosis is a condition that occurs when the body produces excessive quantities of acid or when the kidneys are not removing enough acid from the body. It causes very slight decreases in blood pH and plasma bicarbonate, though if the extent of acidosis is particularly long even a low level of acidosis can be greatly impactful.[2] Acidosis has been linked to high sodium diets, osteoporosis, and includes a risk for muscle wasting and kidney stone formation. The alkaline diet involves the consumption of alkaline-forming foods such as fruits and vegetables and is low in PRAL. This diet has a range of health benefits due to it being high in potassium, magnesium and bicarbonate. Some of the benefits are the advancement in bone mineral density, muscle mass, protection from chronic illnesses, a decrease of tumor-cell invasion and metastasis, and the successful expulsion of toxins from the body.[3]

There are certain health benefits from having an alkaline diet that relate to improving or maintaining the homeostasis pH of the different organs, fluids, and membranes of the body as stated in the article "The Alkaline Diet: Is There Evidence That an Alkaline pH Diet Benefits Health?" written by Gerry K. Schwalfenberg.[4] Schwalfenberg listed the following benefits in his conclusion: (1) Increase in vegetable and fruit intake would improve the K/Na ratio and may benefit bone health, reduce muscle wasting, as well as mitigate other chronic diseases such as hypertension and strokes. (2) The resultant increase in growth hormone with an alkaline diet may improve many outcomes from cardiovascular health to memory and cognition. (3) An increase in intracellular magnesium, which is required for the function of many enzyme systems, is another added benefit of the alkaline diet. Available magnesium, which is required to activate vitamin D, would result in numerous added benefits in the vitamin D apocrine/exocrine systems.

Your food choices affect your intake of potassium, sodium and bicarbonate and sequentially affect your body's ability to maintain its stable pH levels. With either a high sodium intake which can result in acidosis or a balanced intake of all substances leading to a healthy beneficial alkaline diet.

Discussion

In the end what is better for you, an alkaline diet or an acidic diet? Based off of the resources used, I concluded that having an alkaline based diet is better for your overall health and immune system. While the idea that either diet can directly affect your blood pH is incorrect, you can have positive or negative effects on the pH for organs in your body. We know that the alkaline and acidic diets cannot affect the blood pH directly as great changes can be life-threatening, but the diets can affect how much your body has to compensate in order to maintain that pH. Each diet has to do with whether the foods you are ingesting are acid-forming or alkaline-forming. Are the foods acidic once they are consumed or are the products more alkaline? If you have an acid-forming diet you will typically have a high PRAL (Potential Renal Acid Load) which includes the following: animal protein, dairy foods, and refined grains. Having high PRAL over time can lead to a condition called metabolic acidosis which is when the body is producing excess acid or the kidneys are not removing enough acid from the body. This condition can lead to slight decreases in blood pH and plasma bicarbonate. It is proven that even a low form of acidosis overtime can result in osteoporosis and increase the chances of muscle wasting and kidney-stone forming. Now having an alkaline-forming diet has very different result than an acid-forming diet. The alkaline increases the functioning of your body and overall health. It does not affect your blood pH directly but increases your body's ability to maintain its standard pH. The acid-forming diet mainly consists of disadvantages and results in bad condition or illnesses, while the alkaline-forming diet is made up many health benefits to your health, longevity, growth, your body's ability to combat illnesses and much more.

References

Mousa, H. (2016). *Health Effects of Alkaline Diet and Water, Reduction of Digestive-tract Bacterial Load, and Earthing.*
https://www.researchgate.net/publication/301497159_Health_Effects_of_Alkaline_Diet_and_Water_Reduction_of_Digestive-tract_Bacterial_Load_and_Earthing

Welch et al. (2014). "A higher alkaline dietary load is associated with greater indexes of skeletal muscle mass in women".
http://www.twinsuk.ac.uk/wp-content/uploads/2012/11/Welch-Ost-Int-2012.pdf

Kim, B. (2017) *"The Truth About Alkalizing Your Blood"*
http://drbenkim.com/ph-body-blood-foods-acid-alkaline.htm

Foroutan R. (2016) *Alkaline Diet: Does pH Affect Health and Wellness?*
https://foodandnutrition.org/may-june-2016/alkaline-diet-ph-affect-health-wellness/

Lanford J. (2011) *Acidic Foods and Alkaline Foods - Knowing the Difference*
http://www.cancerdietitian.com/2011/10/acidic-foods-and-alkaline-foods-knowing-the-difference.html

Schwalfenberg G. (2011) *The Alkaline Diet: Is There Evidence That an Alkaline pH Benefits Health?*
https://www.ncbi.nlm.nih.gov/pmc/articles/PMC3195546/

Skin Graft Rejection

Patrick Maw

Abstract

Skin Graft rejections can occur seemingly at random, and on a cellular scale appear as molecular warfare. If the patient's skin cells are incompatible with those of the donor, the patient's body will undergo an immune response and release T Cells which is enough to trigger an immune reaction, causing NK Cells to eventually destroy the allogeneic graft. The presence of latent viruses and other factors such as unsanitary conditions can result in the termination of donor skins.

Background

The process of skin grafting is a surgery involving the transplant of skin cells from one source to another. Skin grafts are the appropriate surgery for some injuries where skin is damaged or destroyed. Some skin grafts can be rejected if the donor's skin is not compatible with the patient's skin, which would lead to the death of the donor cells.

Skin grafts can be rejected for reasons such as infection, or if the patient skin graft will not heal to where it was placed. I think that the antibodies produced by the body in response to the missing skin may kill the donors skin if your body has a negative reaction.

In the USA, approximately 2.4 million burn injuries are reported annually. My question is important because many people require this treatment as an appropriate recovery for many injuries and disorders which can result in the deterioration of skin cells. Some of these injuries include flesh eating diseases, car accidents and severe burns. Research on this particular topic can provide valuable insight in regard to the careful process of skin grafting and the consequences of when it goes wrong.

Results

The transplantation of incompatible skin grafts are connected to inflammatory immune responses which eventually leads to the destruction of donor cells followed by the rejection of the skin graft[1]. Several studies demonstrate that NK cells, activated due to missing skin molecules, are involved in allogeneic skin graft rejection via direct killing of donor cells. As a result, current clinical skin transplantation is limited to the grafting of small patches of an individual's own skin.

Recent studies[2] have suggested that T cells can become activated from the recognition of donor MHC molecules (skin cells) transferred on the patient's antigen containing cells, activation of T cells is sufficient to trigger acute rejection of allogeneic skin grafts. In addition, antibodies contribute to the rejection process either by killing allogeneic targets or by targeting donor cells and forming immune complexes.

If a constant blood supply is not maintained after surgery, the skin graft flap may not graft successfully, and could die[3]. Sometimes the blood supply is compromised by swelling around the graft site. Graft failure may also occur is the skin rejected by the patient's body. Infection is common and grafts are easily destroyed. In cases of infection, oozing, redness, itching and pain occur.

Discussion

Many patients with major burns involving at least 25% of their total body surface die. The transplantation of large patches of skin would save their lives but allogeneic skin grafts are commonly rejected. Skin grafting is a careful process, that even if accomplished perfectly can trigger a potent reaction by the host's immune system causing rapid degradation of donor cells and rejection of the graft. This is due to an immune response that is the result of the antigens on the skin's surface triggering NK cells to destroy the donors skin as a perceived threat. With the knowledge of these reactions, transplants have mainly been from the patient's own body, to avoid an incompatible graph.

References

[1] Benichou et al. (2010.) "Immune recognition and rejection of allogeneic skin grafts" https://www.ncbi.nlm.nih.gov/pmc/articles/PMC3738014/

[2] (2011) Benichou G "Immune recognition and rejection of allogeneic skin grafts." https://www.ncbi.nlm.nih.gov/pubmed/21668313

[3] (2011) Veach, M "Skin Graft Complications" https://www.livestrong.com/article/53266-skin-graft-complications/

Reversal of the Aging Process

CJ James

Abstract

Some scientists are trying to determine if there are genes that regulate aging. Other scientists have been discovering anti-aging mechanisms in animals including calorie restriction and the chemical resveratrol found in red wine. Scientists believe that these diverse strategies work, suggesting there may be more than one way to reverse aging. One scientist says "that multiple complementary therapies may be required to significantly extend longevity". If scientists can apply these strategies to humans, then maybe one day people could live generations longer.

Background

Is it possible to reverse aging? I've seen reverse aging in movies and TV shows. When you reverse aging, it keeps you looking young as if you were immortal. In two of my anime shows I watch that are called Inuyasha and Naruto, aging reversal is depicted. In Inuyasha, reverse aging is when there are demons or half demons who can take human form and can live up to 1000 years or more. In Naruto there is a character named Hidan who is immortal and doesn't age. In Naruto the main character has a tailed beast/demon that doesn't keep him from aging, however it helps him heal faster than most humans and gives him extensive power. I don't predict that age reversal will be possible because demons aren't real. My inquiry question is important because if it was possible to reverse aging then there would be more people polluting the earth and destroying it.

Methods

I was able to get my information from websites that give me detailed information. I was able to find three good websites that have the information that I needed.

Results

Scientists have proven that it is possible to slow or reverse aging in mice by undoing changes in gene activity associated with aging. By exploiting genes to turn adult cells back into embryonic-like cells, researchers at SIBS[3] (Salk Institute for Biological Studies) reversed the aging of mouse and human cells in vitro. S.I.B.S extended the life of a mouse with the an accelerated-aging condition and successfully promoted recovery from an injury in a middle-aged mouse. This study adds weight to the scientific argument that aging is largely a process of epigenetic changes, alterations outside of the genetic code that make genes more active. Over the course of life, cell activity regulators get added to or removed from genes. These changes in humans can be caused by smoking, pollution or the environment factors which dials the gene's activities up or down.

As changes accumulate our muscles weaken, our mind slows down and we become vulnerable to disease.[1] In my last resource they were also experimenting on mice. Their team was able to find out that giving older mice a chemical called NAD for just one week made a 2-year old mice tissue resemble that of 6-month old mice. When mammals age, the levels of NAD drop by 50%. With less NAD, communication between the cell and its mitochondrial energy source falters and the cell becomes vulnerable to aging. By giving your cell adequate amounts of NAD, aging can theoretically be reversed.

Red wine also has anti-aging effects. The compound resveratrol has been show to reverse the effects of aging in cells. However, scientists aren't yet ready to say that resveratrol could lead to immortal cells.[2]

Discussion

Scientists have shown that it is possible to reverse aging but they haven't yet figured out if it work in humans. They have tested their work on mice and have proven it to be successful. The only chemicals that have been shown to help with reverse aging for humans are found in red wine. I believe that scientists will eventually be able to reverse aging. I wouldn't want to have them test it on humans though because what if something happens to them and it isn't reversible? If scientists do make it happen they should be ready to make everyone live longer. I know that there are scientists who are working to reverse aging, but I think that if they can do it in mice then they should be able to do it to humans.

References

[1] Weintraub, K (2016). "Aging Is Reversible-At Least in Human Cells and Live Mice". https://www.scientificamerican.com/article/aging-is-reversible-at-least-in-human-cells-and-live-mice/

[2] Park, A (2013). "Reversing Aging: Not as Crazy as You Think". http://healthland.time.com/2013/12/19/reversing-aging-not-as-crazy-as-you-think/

[3] Ocampo, A. et al. (2016). "In vivo amelioration of age-associated hallmarks by partial reprogramming". https://www.salk.edu/news-release/turning-back-time-salk-scientists-reverse-signs-aging/

www.ingramcontent.com/pod-product-compliance
Lightning Source LLC
Chambersburg PA
CBHW051105180526
45172CB00002B/787